Indexing

Indexing

A Nuts-and-Bolts Guide
for Technical Writers

Kurt Ament

William Andrew Publishing

Norwich, NewYork, U.S.A.

William Andrew
publishing

President and CEO: William Woishnis
Vice President and Publisher: Dudley R. Kay
Production Manager: Kathy Breed

Production Services: **TIPS** Technical Publishing, Inc.
Copy Editor: Jeannine Kolbush
Design and Composition: Robert Kern
Proofreader: Darryl Hamson
Cover Designer: Brent Beckley

Published in the United States of America by
William Andrew Publishing
13 Eaton Avenue
Norwich, New York 13815
1-800-932-7045
www.williamandrew.com

ISBN: 0-8155-1481-6
Printed in the United States of America

10 9 8 7 6 5 4 3 2 1

This book may be purchased in quantity discounts for educational, business, or sales promotional use by contacting the Publisher.

Trademarks

Adobe® is a registered trademark of Adobe Systems Incorporated. Acrobat™, FrameMaker™, and Photoshop™ are trademarks of Adobe Systems Incorporated. Apple®, iMac®, and Macintosh® are registered trademarks of Apple Corporation. CINDEX™ is a trademark of Indexing Research. HP® is a registered trademark of Hewlett-Packard Company. IBM® is a registered trademark of International Business Machines, Inc. LaserJet™ is a trademark of Hewlett-Packard Company. Microsoft®, Internet Explorer®, Windows®, and Word® are registered trademarks of Microsoft Corporation. Netscape™, Netscape Messenger™, and Netscape Navigator™ are trademarks of Netscape Communications Corporation. Norton AntiVirus™ and Norton Utilities™ are trademarks of Symantec Corporation. UNIX® is a registered trademark of the Open Group. Other product names mentioned in this guide may be trademarks or registered trademarks of their respective companies and are hereby acknowledged.

C O N T E N T S

About this guide   vii

Audience viii
Mission viii
Strategy ix
Tactics ix

1 About indexing   1

Anyone can index 2
    Indexers are made, not born 2
    Indexing does not have to be painful 2
    Nothing succeeds like success 2
Sequential indexing 3
    Separating content from format 3
    Keeping your sanity 3
Usable indexes increase profits 4
    Usable indexes improve documentation 4
    Usable documentation improves products 4
    Usable products sell 5

2 Start indexing   7

Step 1 Index chapters 8
Step 2 Index procedures 9
Step 3 Index topics 10
Step 4 Index product names 11
Step 5 Index product components 12
Step 6 Index front and back matter 13

Step 7 Edit your index.. 14

Step 8 Create "see" references.. 15

Step 9 Create "see also" references 16

Step 10 Test your index... 17

3. Indexing guidelines   19

Abbreviations 20

Acronyms 23

Articles 24

Back matter 26

Capitalization 28

Cross-references 31

Front matter 34

Interface components 36

Keyboard shortcuts 39

Master indexing 41

Nesting 46

Page ranges 49

Prepositions 52

Procedures 54

Product names 57

Scheduling 60

"See" references 63

"See also" references 67

Sorting 72

System messages 77

Tools 80

Topics 82

Index   87

About this guide

This guide explains in plain language and by example exactly how to index any type of print or online documentation quickly, easily, and effectively. The sequential indexing method presented in this guide has been battle-tested in high-pressure publishing organizations in a variety of high-tech industries over the space of a decade. Because it is based exclusively on real-world success, this indexing method is bulletproof.

Audience

This guide is designed for anyone who needs to index technical publications:

- **Technical writers and editors**

  This guide benefits professional writers and editors faced with the often daunting task of indexing complex print and online publications.

- **Technical publications managers**

  This guide benefits technical publications managers who want to standardize corporate publications in a way that improves the quality of print and online documentation while saving time and money.

- **Teachers and students of technical communication**

  This guide benefits teachers and students of technical communication at universities, technical colleges, and trade schools who want to learn and keep up with industry standards for technical indexing.

Mission

This guide has two integrated goals:

- **Immediate success**

  The primary goal of this guide is to teach anyone how to index any type of technical publication in a very short time and with minimal effort.

- **Continuing success**

  The secondary goal of this guide is to provide a universal indexing strategy that any publishing organization can apply to any type of technical publication in any medium.

Strategy

To achieve immediate and continuing success, this guide delivers three distinct benefits:

- **Usability**
  This guide is designed to be scanned, not read cover to cover. It employs user-centered information mapping techniques that make it easy for a wide variety of readers to find exactly the level of information they need, when and where they need it.

- **Reusability**
  This guide presents a highly structured indexing method that can be customized by publishing departments to meet the specific needs of their companies and products.

- **Reliability**
  This guide is a success story. It presents a success-based indexing method developed, implemented, and refined in publishing departments at diverse multinational corporations over the space of a decade.

Tactics

To achieve usability, reusability, and reliability, this guide explains indexing on three levels:

1 **"About indexing"**
  This chapter introduces the sequential indexing method. A structural overview of products, product documentation, and document indexes explains why the sequential indexing method is so successful.

2 **"Start indexing"**
  This chapter contains sequential step-by-step procedures. These procedures explain exactly how to index any type of document, from start to finish. Each procedure step contains cross-references to detailed indexing guidelines in Chapter 3, "Indexing guidelines."

3   **"Indexing guidelines"**
This chapter contains detailed indexing guidelines. These guidelines explain everything from indexing acronyms to creating "see also" references to building master indexes. The guidelines are sorted alphabetically for easy reference. Each guideline contains positive and negative examples as well as cross-references to related guidelines and procedures.

This guide practices what it preaches. It is fully indexed, following its own guidelines.

TIP   Although this guide uses the model of printed software guides to explain sequential indexing, you can easily adapt this method to index online help systems, product specifications, reference manuals, technical libraries, and websites.

1

About indexing

Indexing is not rocket science. Indexing is a relatively simple task that can be learned quickly and painlessly. All you need to start indexing successfully is a practical plan of attack. This chapter outlines a practical indexing method that has proven itself in real-world technical publishing environments. If you follow this method, you will succeed as an indexer.

Anyone can index

Indexing is a task performed every day by perfectly normal people like you.

Indexers are made, not born

In the technical publishing field, indexers are technical writers and editors who, at some point in their careers, have an indexing assignment thrust upon them. Typically the orders come from someone further up the corporate food chain. The person giving the orders knows indexes are important but does not have a clear idea what an index is, let alone how to build one. The person receiving the orders is left alone in the dark. And so begins the long, lonely march into the unknown.

Indexing does not have to be painful

As a novice indexer, you don't need to reinvent the wheel or blaze new trails. Many indexers have indexed before you. These pioneers have learned, often the hard way, what works and what doesn't. In the process, they have laid the groundwork for your success as an indexer.

Nothing succeeds like success

To succeed as an indexer, simply follow the indexing method used by successful indexers. The most successful indexing method used by professional indexers is quick, easy, and foolproof. This method is called sequential indexing.

Sequential indexing

Sequential indexing is a proven method for creating indexes. This indexing method *separates index creation (content) from index editing (format)*. By separating content from format, the sequential method breaks an otherwise overwhelming task into bite-sized tasks anyone can perform under real-world conditions. By performing these simple tasks one step at a time, you will build accurate and usable indexes quickly and easily.

TIP   Although sequential indexing has its origin in printed software guides, you can easily adapt the method to any medium (for example, online help systems or websites).

Separating content from format

Like all successful documentation methods, sequential indexing separates content from format:

1   **Create index content**
    In the early stages of developing your index, you *focus exclusively on the content of the document you are indexing.* You index distinct types of information, type by type, without regard to the format of the index itself. By ignoring the format of the index, you create factually accurate index entries with minimal effort.

2   **Edit index format**
    Once you have created your index entries, you *focus exclusively on the format of the index.* When editing the format of the index, you concentrate on usability alone. That is, you tailor the format of the index to the specific needs of your users. This exclusive focus on usability is the best method for transforming raw content into a truly usable index.

Keeping your sanity

Indexing one step at a time breaks your work into easily manageable tasks. This task-driven approach insulates you from a chronic problem in most work environments: chaos. Sequential indexing assumes you will be interrupted frequently. While indexing, you will be interrupted by telephone calls, emergency meetings, even computer failure. By focusing on simple tasks, you will recover from these interruptions with very little effort.

Usable indexes increase profits

Sequential indexing does more than make your job easier. It enables you to create usable indexes. Usable indexes improve product documentation. Usable product documentation improves products. And usable products sell.

Usable indexes improve documentation

An index is a documentation tool that saves time and energy. This tool enables users to find information in documentation quickly and easily. Rapid information retrieval is critical to the success of product documentation. Unlike classical literature or pulp fiction, product documentation is almost always scanned, not read cover to cover. In fact, users often turn to product documentation as a last resort, particularly when something has gone horribly wrong with the product they are using. The less time and energy it takes users to find the information they need, the more usable the documentation.

The more usable the documentation, the easier it is to index. Well-structured documents are easy to index, and poorly structured documents are difficult to index. If a document is difficult to index, the document most likely has structural problems. That is the bad news. The good news is that the process of indexing uncovers these otherwise hidden structural problems. For this reason, *professional editors routinely use the process of indexing to diagnose and correct structural problems within product documentation.*

Usable documentation improves products

In the same way that usable documentation is easy to index, usable products are easy to document. The converse is also true: unusable products are difficult to document. If a product is difficult to document, it is very likely that the product has structural problems. Many of these structural problems are uncovered during the documentation process itself. For this reason, *product development teams routinely use the process of documentation to diagnose and correct structural problems within products.* In other words, documentation serves as product quality assurance.

The dual role of product documentation is underlined by an overworked joke in the software industry. Software engineers often refer to program errors, or bugs, as "undocumented features." This dark humor is not shared by product users. As the "first users" of the products they document, *technical writers routinely uncover product errors*. In well-functioning product development groups, *errors uncovered by technical writers are corrected by product developers*. Obviously, these corrections make products easier to use.

Usable products sell

Products that are easy to use attract and retain customers. This fact is indisputable. In a free market, customer loyalty depends on usable products. Usable products depend on usable product documentation, which in turn depends on usable indexes. The better the index, the better the documentation. The better the documentation, the better the product. The better the product, the better the sales. In other words, *usable indexes contribute directly to product sales.*

2

Start indexing

This chapter shows you how to index any document, from start to finish, step by step. Each step is cross-referenced to detailed guidelines in Chapter 3, "Indexing guidelines."

Step 1   Index chapters 8

Step 2   Index procedures... 9

Step 3   Index topics 10

Step 4   Index product names 11

Step 5   Index product components 12

Step 6   Index front and back matter 13

Step 7   Edit your index 14

Step 8   Create "see" references 15

Step 9   Create "see also" references............................. 16

Step 10   Test your index 17

Step 1   **Index chapters**

Read this

The largest and most important components of any book are its chapters. Chapters are labeled with chapter numbers and titles. For example, the title of Chapter 2, the chapter you are currently reading, is "Start indexing."

To start your index, begin with the chapters. Index each chapter of the book. Do not index non-numbered sections of the book, such as the title page, copyright page, preface, introduction, appendices, or glossary.

Before indexing a chapter, determine whether the chapter as a whole contains narrative topics or step-by-step procedures:

- **Topics**
  If the chapter contains any narrative text (for example, "About Adobe FrameMaker+SGML"), index the chapter as a topic.

- **Procedures**
  If the chapter contains nothing but step-by-step procedures (for example, "Installing Adobe FrameMaker+SGML"), index the chapter as a procedure.

Do this

To index a chapter, follow these steps:

Steps	Guidelines
1 Decide whether the chapter is procedural or topical.	▪ *Procedures* on page 54 ▪ *Topics* on page 82
2 Create an index entry for the chapter.	▪ *Procedures* on page 54 ▪ *Topics* on page 82
3 If the chapter is longer than one page, create a page range for the entry.	▪ *Page ranges* on page 49

Step 2   **Index procedures**

Read this

Procedures tell users *how* to perform tasks, step by step.

Well-designed procedures are:

- **Clearly labeled**
  Headings indicate a specific action (for example, "To print a document").

- **Explicitly numbered**
  Each procedure step is explicitly numbered and followed immediately by explicit instructions, not narrative text.

Create index entries for each procedure in the book. Work your way through the book as it is written, chapter by chapter.

TIP   Some procedures contain subprocedures. When indexing procedures, ignore these hierarchies. Simply index each procedure and subprocedure as a stand-alone entry. Later, when editing the index, you will "nest" procedure entries into hierarchies that apply to the book as a whole (see *Edit your index* on page 14).

Do this

To index a procedure, follow these steps:

Steps	Guidelines
1 Create an entry for the procedure.	▪ *Procedures* on page 54
2 If the procedure is longer than one page, create a page range for the entry.	▪ *Page ranges* on page 49

Step 3   Index topics

Read this

Topics are narrative texts that answer one or more of the following questions:

- Who?
- What?
- When?
- Where?
- Why?

If they are well designed, topics are clearly labeled with headings that indicate exactly what the text contains (for example, "Types of computer viruses"). Create index entries for each distinct topic in each chapter of the book.

TIP   Some topics contain subtopics. When indexing topics, ignore these hierarchies. Simply index each topic and subtopic as a stand-alone entry. Later, when editing the index, you will "nest" topical entries into hierarchies that apply to the book as a whole (see *Edit your index* on page 14).

Do this

To index a topic, follow these steps:

Steps	Guidelines
1 Create an entry for the topic.	- *Topics* on page 82
2 If the topic is longer than one page, create a page range for the entry.	- *Page ranges* on page 49

Step 4   Index product names

Read this

Product names are trademarked names for products. For legal reasons, product names should be indexed exactly as they are trademarked.

Index product names based on how they are used in the book:

- **Primary products**
  If a product name is the subject of the book, it is implied throughout the index. Do not include this product name as an index entry.

- **Secondary products**
  If the product name is *not* the subject of the book, include both the trademarked name and the name of the product's manufacturer (for example, "IBM AS/400").

- **Relevant discussions**
  Index only meaningful discussions of the product, not every mention of the product.

Do this

To index a product name, follow these steps:

Steps	Guidelines
1 Create an entry for the product name.	- *Product names* on page 57
2 If the discussion of the product is longer than one page, create a page range for the entry.	- *Page ranges* on page 49

Step 5   **Index product components**

Read this

In computer programs, product components are the parts of the product with which users interact:

- **Interface components**

  Interface components are the visible or audible parts of a computer program. They make up the user experience.

- **Keyboard shortcuts**

  Keyboard shortcuts enable users to execute program commands from the keyboard rather than with the mouse.

- **System messages**

  System messages are words, phrases, and sentences delivered by a program to users in particular situations.

Each type of product component is indexed differently.

Do this

To index product components, follow these steps:

Steps	Guidelines
1 Create an entry for each interface component in the book.	▪ *Interface components* on page 36
2 Create an entry for each keyboard shortcut in the book.	▪ *Keyboard shortcuts* on page 39
3 Create an entry for each system message in the book.	▪ *System messages* on page 77

Step 6   Index front and back matter

Read this

Front and back matter are the parts of a book before and after the chapters:

- **Front matter**

  Front matter appears *before* the chapters of the book:

  > Title page
  > Copyright page
  > Table of contents
  > Preface
  > Introduction

- **Back matter**

  Back matter appears *after* the chapters of the book:

  > Appendices
  > Glossary
  > Index

Do not index front and back matter themselves (for example, "Appendix A"). Create entries for critical information contained in front and back matter (for example, "troubleshooting").

Do this

To index front and back matter, follow these steps:

Steps	Guidelines
1 Create an entry for each important topic in the preface or introduction to the book.	▪ *Front matter* on page 34
2 Create an entry for each important topic or procedure in the appendices to the book.	▪ *Back matter* on page 26

Step 7   **Edit your index**

Read this

Editing an index is like editing any other type of document. When editing documents, you focus primarily on the usability of the documents themselves, not on the products that have been documented. When editing your index, you focus on the usability of your index, not on the book you have indexed.

The task of editing an index is completely distinct from creating an index. It requires a completely different mindset. To edit your index successfully, you need to switch mental gears. Begin editing only after you have created all of your index entries.

TIP   In the same way that you should never throw good money after bad money, you should never add cross-references to a poorly structured index. Always edit your index *before* you cross-reference it with "see" and "see also" references. By cleaning up your index entries first, you provide a stable structure for your cross-references. This stable structure ensures that you add cross-references only once.

Do this

To edit your index, follow these steps:

Steps	Guidelines
1 Remove all articles.	▪ *Articles* on page 24
2 Remove all prepositions.	▪ *Prepositions* on page 52
3 Make sure index entries are sorted properly.	▪ *Sorting* on page 72
4 Make sure index entries are nested properly.	▪ *Nesting* on page 46

Step 8 Create "see" references

Read this

A "see" reference is a cross-reference from an alias to a preferred word or phrase.

Example:

> Microsoft Word, 13
> WinWord. *See* Microsoft Word

In this example, the preferred word or phrase includes a page number ("Microsoft Word, 13"). The alias does not include a page number. It has a "see" reference pointing to the preferred word or phrase ("WinWord. *See* Microsoft Word").

Aliases with "see" references enable users to find information quickly, using whatever keywords come to mind during their search. At the same time, "see" references enable you to index items only once, thereby avoiding redundant work.

Do this

To create "see" references, follow these steps:

Steps	Guidelines
1 Create "see" references for all abbreviations and acronyms.	▪ *Abbreviations* on page 20 ▪ *Acronyms* on page 23
2 Create "see" references for alias entries.	▪ *"See" references* on page 63
3 Test all "see" references.	▪ *"See" references* on page 63

Step 9 Create "see also" references

Read this

A "see also" reference is a cross-reference from one index item to a related index item:

deleting files
    See also recovering files
recovering files
    See also deleting files

Although they are valid by themselves, "see also" references, like shoes, are meant to function in pairs. "See also" references identify index entries that are closely related to each other. It is highly probable that users who see one index entry would also like to see the other entry, and vice versa.

Unlike "see" references, "see also" references are not essential to the usability of an index. Nevertheless "see also" references underline relationships between product features, functions, and procedures that would not otherwise be obvious to users. By adding "see also" references, you give a certain intelligence to an index.

Do this

To create "see also" references, follow these steps:

Steps	Guidelines
1 Add reciprocal "see also" references.	▪ *"See also" references* on page 67
2 Test all "see also" references.	▪ *"See also" references* on page 67

Step 10 **Test your index**

Read this

Now comes the moment of truth. It is time to test your index.

There are three ways to test an index:

- **Right-brain test (quick)**
  The quickest way to test your index is to print it out, turn it upside down, hold it at a 90-degree angle from your body, and squint. If you see well-balanced columns and geometrical shapes, your index is well structured. If you don't, there is something wrong with the structure.

- **Left-brain test (slow)**
  A not-so-quick way to test your index is to systematically check all page numbers and cross-references for accuracy. If you are using a publishing system that embeds index markers in the body of your document, you need only to spot-check your entries. If not, find people lower on the totem pole to find your mistakes. In an effort to make you look bad, they will test your index with a vengeance.

- **User test (rare)**
  The best way to test your index is to give it to end users, or their functional equivalents. Hand users a checklist of tasks that take them into the bowels of your product. Ask them to use the index to find the information they need. If your index has problems, you will definitely hear about it.

Do this

To test your index, follow these steps:

Steps

1	Conduct a left-brain test. Correct your index as needed.
2	Conduct a right-brain test. Correct your index as needed.
3	If you have the resources, conduct a user test. Correct your index as needed.

3

Indexing guidelines

This chapter contains detailed guidelines to indexing everything from acronyms to master indexes to "see also" tags. Each guideline contains positive and negative examples.

Abbreviations 20
Acronyms 23
Articles 24
Back matter.. 26
Capitalization.. 28
Cross-references 31
Front matter... 34
Interface components.. 36
Keyboard shortcuts 39
Master indexing 41
Nesting.. 46
Page ranges 49
Prepositions 52
Procedures 54
Product names 57
Scheduling 60
"See" references 63
"See also" references... 67
System messages 77
Tools 80
Topics 82

Abbreviations

See also

For related guidelines, see also:

- *Acronyms* on page 23
- *Product names* on page 57
- *"See" references* on page 63

Read this

Abbreviations are shortened forms of words and phrases. These linguistic shortcuts make excellent index entries because they enable users to access information fast.

Within an index, the speed with which an abbreviation is recognized is based on its familiarity. If an abbreviation is well known, most users will look in the index for the abbreviation rather than the word or phrase it abbreviates. If the abbreviation is not well known, most users will look in the index for the plain-language word or phrase rather than the abbreviation.

Types of abbreviations

There are three kinds of abbreviations:

- **Common abbreviations**
  Common abbreviations (for example, "HTML") are more familiar to users than the words they abbreviate (for example, "HyperText Markup Language").

- **Uncommon abbreviations**
  Uncommon abbreviations (for example, "UMB") are less familiar to many users than the words they abbreviate (for example, "upper memory block").

- **Trademarked abbreviations**
  Trademarked abbreviations (for example, "IBM") are legally shortened names for companies or products (for example, "International Business Machines, Inc."). These proprietary abbreviations are usually more familiar to users than the names they abbreviate.

Imposing predictability on abbreviations

Given the diversity of abbreviations, it is difficult to predict which abbreviations users will look for and which abbreviations they have never heard of. This unpredictability is compounded, of course, by the diversity of users themselves. For this reason, it is wise to impose predictability on abbreviations and users alike. That is, pick a single, sensible strategy for indexing abbreviations and stick with it throughout your index.

The best strategy for indexing abbreviations is to index the abbreviations and the words they abbreviate. To save yourself from redundant work—not a trivial matter when page numbers or hyperlink locations change—list abbreviations as normal entries. Then set up full-length entries as cross-references ("see" references) pointing back to the abbreviations.

Do this

When indexing abbreviations, follow these guidelines:

- **Index common abbreviations**
  Index abbreviations rather than the words they abbreviate. Use "see" references to direct users from the complete words to their abbreviations.

Bad	Good
Extensible Markup Language, 3	Extensible Markup Language. *See* XML XML, 3
Portable Document Format, 4	PDF, 4 Portable Document Format. *See* PDF
Standard Generalized Markup Language, 5	SGML, 5 Standard Generalized Markup Language. *See* SGML

- **Index uncommon abbreviations**

  Index uncommon abbreviations rather than the words they abbreviate. Use "see" references to direct users from complete words to their abbreviations.

Bad	Good
Cyclical Redundancy Checking, 6	CRC, 6 Cyclical Redundancy Checking. *See* CRC
Data Encryption Standard, 5	Data Encryption Standard. *See* DES DES, 5
upper memory block, 3	UMB, 3 upper memory block. *See* UMB

- **Index trademarked abbreviations**

  Index trademarked abbreviations rather than the words they abbreviate. Use "see" references to direct users from complete words to their abbreviations.

Bad	Good
International Business Machines, 4	IBM, 4 International Business Machines. *See* IBM
Institute of Electrical and Electronics Engineers, 15	IEEE, 15 Institute of Electrical and Electronics Engineers. *See* IEEE

Acronyms

See also

For related guidelines, see also:

- *Abbreviations* on page 20
- *"See" references* on page 63

Read this

Acronyms are abbreviations that can be pronounced as words. These pronounceable abbreviations are usually more familiar to users than the more cumbersome words they abbreviate. Always index acronyms rather than the words they abbreviate.

Do this

When indexing common acronyms, follow these guidelines:

- **Index acronyms**
  Index acronyms rather than the words they abbreviate.

Bad	Good
American Standard Code for Information Interchange, 2	ASCII, 2

- **Cross-reference acronyms**
  Use "see" references to direct users from complete words to their acronyms.

Bad	Good
BASIC, 3	BASIC, 3 Beginners All-purpose Symbolic Instruction Code. *See* BASIC

Articles

See also

For related guidelines, see also:

- *Prepositions* on page 52

Read this

Articles are the three most common adjectives:

- a

- an

- the

Like all adjectives, articles modify nouns by either limiting them or making them more precise.

Types of articles

There are two kinds of articles:

- **Indefinite articles: "a" and "an"**

  Indefinite articles denote generic nouns. For example, "a computer program" denotes any program, not a specific program.

- **Definite articles: "the"**

  Definite articles denote specific nouns. For example, "the computer program" denotes a specific program, not any program.

Articles are essential to sentences because they indicate whether the nouns are specific things or just a group of similar things. Articles clarify the meaning of sentences and make reading easier.

Articles are deadly to indexes

Although they are essential to sentences, articles are deadly to indexes because they take up physical space without adding any real value. Articles do not clarify the meaning of index entries or make indexes easier to use.

To understand why articles are counter-productive in indexes, you need to understand three simple facts about indexes:

- **Index entries are catch-phrases**

  Index entries are simple words designed to express the greatest possible meaning in the smallest possible space. Like advertising slogans, index entries emphasize the essential (nouns, verbs) at the expense of everything else (adjectives, adverbs, prepositions).

- **Indexes are organized alphabetically**

  Because index entries are ordered alphabetically, it is essential that the first word of each entry convey as much information as possible. The only parts of speech that convey stand-alone information are nouns and verbs. Of course, the best way to destroy an index would be to begin entries with "a," "an," or "the."

- **Users scan rather than read indexes**

  An index contains nothing but simple words (and their associated page numbers or hyperlinks) that are organized alphabetically. In other words, an index is a glorified list. In the same way that no one reads a telephone book cover to cover, no one reads an index from beginning to end. Instead, users scan indexes for useful information. In this context, articles are completely useless.

Do this

When indexing any entry, follow this guideline:

- **Avoid articles**

  Do not use articles (a, an, the) in index entries.

Bad	Good
a network cable, 4	network cable, 4
printing a document, 3	printing document, 3
the Windows Explorer, 5	Windows Explorer, 5

Back matter

See also

For related guidelines, see also:

- *Front matter* on page 34

Read this

Back matter is reference material that appears *after* the chapters of a book:

- Appendices
- Glossary
- Index

These reference sections are not part of the book content and therefore should not be indexed as such.

Appendices can contain critical reference information such as:

- Emergency procedures
- System messages
- Troubleshooting procedures

This critical reference information should be indexed.

TIP   If you find yourself repeatedly tempted to index an entire appendix because it contains information essential to understanding the subject of the book, consider converting the appendix into a chapter. The section has most likely been misclassified as back matter (an appendix) and should be core content (a chapter).

Do this

When indexing back matter, follow these guidelines:

- **Do not index back matter as such**
  Do not create index entries for appendices, glossaries, or indexes themselves.

Bad	Good
glossary, 101	—
index, 123	—

- **Index reference information in back matter**
  List critical reference information, such as emergency procedures, system messages, and troubleshooting procedures, that appears in back matter.

Bad	Good
Appendix A. System messages, 87	system messages, 87–88
Appendix B. Emergency data recovery procedures, 89	emergency data recovery procedures, 89–94
Appendix C. Troubleshooting, 95	troubleshooting, 95–99

Capitalization

See also

For related guidelines, see also:

- *Product names* on page 57
- *"See" references* on page 63

Read this

Proper capitalization is critical to the usability of an index. Indexes are lists, and lists are scanned, not read. To succeed, index entries must be clear, concise, and to the point. There literally is no time or space for long-winded explanations. Visual (pre-verbal) communication is critical. How an index entry is capitalized signals in the broadest possible terms what type of information the entry contains.

Index entries can be capitalized in one of three ways:

- **Lowercase (abc)**

  The clearest signal for garden-variety entries is lowercase (for example, "printing documents").

- **Uppercase (ABC)**

  The clearest signal for literal strings, such as computer commands and filenames, is uppercase (for example, "AUTOEXEC.BAT").

- **Initial capitalization (Abc)**

  The clearest signal for interface elements, such as computer program names and menu items, is initial capitalization (for example, "Print Preview").

To avoid confusing these three distinct signals, never capitalize the first word of an index entry just because it is the first word (for example, "Printing documents"). Although this traditional type of capitalization, known as downstyle capitalization, is ideal for document headings, it confuses the visual (pre-verbal) signals of the most carefully constructed index.

Do this

When developing any index entry, follow these guidelines:

- **Use lowercase as your default**
  Use lowercase for all words that are not proper nouns.

Bad	Good
Software applications, 6	software applications, 6
Troubleshooting, 4	troubleshooting, 4

- **Use uppercase for literal strings**
  Use uppercase for literal strings, such as computer commands and filenames, even if these nouns are capitalized differently in the document you are indexing.

Bad	Good
Autoexec.bat, 1	AUTOEXEC.BAT, 1
config.sys, 2	CONFIG.SYS, 2

- **Use initial capitalization for product components**
  Use initial capitalization for products, product components, and interface elements.

Bad	Good
norton utilities, 5	Norton Utilities, 5
file menu, 3	File menu, 3

- **Follow trademarked capitalization**

  Never change the capitalization of trademarked product or component names.

Bad	Good
DBASE files, 1	dBASE files, 1
Imac, 17	Apple iMac, 17

Cross-references

See also

For related guidelines, see also:

- *"See" references* on page 63
- *"See also" references* on page 67

Read this

Indexes contain two distinct types of cross-references:

- **"See" references (nonreciprocal)**
  "See" references are nonreciprocal cross-references from alias index entries to preferred index entries.

- **"See also" references (reciprocal)**
  "See also" references are reciprocal cross-references from index entries to other index entries to which they are related.

The best way to understand "see" and "see also" references is to compare them:

"See" references	"See also" references
Appear alone	Appear in pairs
Point one way	Point two ways
Point to preferred wording	Point to related entries
Never have page numbers	Always have page numbers

"See" references	"See also" references
Cannot have subentries	Can have subentries

SCSI, 25
small computers systems interface.
    See SCSI
	de-installing
	    See also installing
	    HP-UX, 31-32
	    Solaris, 33-34
	installing
	    See also de-installing
	    HP-UX, 27-28
	    Solaris, 29-30

Do this

When developing cross-references, follow these guidelines:

- **Use "see" references to point to preferred wording**

  Use "see" references to point from alias index entries to preferred index entries.

Bad	Good

small computer systems interface
    (SCSI), 13
	SCSI, 13
	small computer systems interface.
	    See SCSI

exit, 5 quit, 5	exit, 5 quit. *See* exit

- **Use "see also" references to point to related entries**

  Use "see also" references to point from index entries to related index entries, and vice versa.

Bad	Good

de-installing
    DEC Alpha NT, 17-18
    Linux, 19-20
installing
    DEC Alpha NT, 13-14
    Linux, 15-16
	de-installing
	    See also installing
	    DEC Alpha NT, 17-18
	    Linux, 19-20
	installing
	    See also de-installing
	    DEC Alpha NT, 13-14
	    Linux, 15-16

Front matter

See also

For related guidelines, see also:

- *Back matter* on page 26

Read this

Front matter is reference material that appears *before* the chapters of a book:

- Title page
- Copyright page
- Table of contents
- Preface
- Introduction

These reference sections are not part of the book content and should not be indexed as such.

Front matter can contain critical reference information such as:

- Document conventions
- System requirements

This critical reference information should be indexed.

TIP   If you find yourself repeatedly tempted to index a front matter section because it contains information essential to understanding the subject of the book, consider converting the section into a chapter. The section has most likely been misclassified as front matter and should be core content.

Do this

When indexing front matter, follow these guidelines:

- **Do not index front matter as such**
  Do not index title pages, copyright pages, tables of contents, prefaces, or introductions themselves.

Bad	Good
copyrights, ii	—
trademarks, ii	—

- **Index reference information in front matter**
  List critical reference information and major topics that appear in prefaces and introductions.

Bad	Good
introduction, ix	system requirements, ix
preface, vii	document conventions, vii

Interface components

See also

For related guidelines, see also:

- *"See" references* on page 63
- *System messages* on page 77

Read this

Interface components are the visible or audible parts of a computer program. These visible or audible parts make up the user experience. That is, users interact directly with interface components.

As a general rule, programs have three types of interface components:

- **Active components (few)**

  Active components are used to *directly run a specific program* (for example, files, menu commands, typed commands, and utilities). These components should be indexed *by name and type*.

- **Passive components (many)**

  Passive components are used to *indirectly run a specific program* (for example, buttons, command parameters, dialog boxes, disks, menus, and windows). These components should be indexed *by type*.

- **Generic components (most)**

  Generic components are used to *directly or indirectly run any program* (for example, combination boxes, drop-down list boxes, icons, list boxes, option buttons, options, pop-up menus, and text boxes). These components should *not* be indexed.

A small number of components actively help users and should therefore be indexed heavily. A larger number of components help the active components and should therefore be indexed lightly. And the vast majority of components are too generic to merit indexing at all.

Do this

When indexing interface components, follow these guidelines:

- **Index active components by name and type**

  Index the following components by name and type:

  Files

  Menu commands (if fewer than 12)

  Typed commands (if fewer than 12)

  Utilities

Bad	Good

commands, 1, 15	commands
	    Delete, 15
	    Open, 1
	Delete command, 15
	Open command, 1

files, 6, 9	AUTOEXEC.BAT file, 9
	.BIN file, 6
	files
	    AUTOEXEC.BAT, 9
	    .BIN, 6

- **Index passive components by name**

  Index the following components by name only:

  Buttons

  Dialog boxes

  Disks

  Menus

  Menu commands (if 12 or more)

  Parameters

  Typed commands (if 12 or more)

  Windows

Use "see" references to point from component type to component name.

Bad	Good
menus, 158, 199	menus. *See entries of specific menus* Norton Partition menu, 158 Wipe Info menu, 199
parameters, 25, 26	HEIGHT parameter, 25 parameters. *See entries of specific parameters* WIDTH parameter, 26
windows, 6, 37	Backup window, 6 Print window, 37 windows. *See entries of specific windows*

- **Do not index generic components**

  Do not index the following components:

  Combination boxes
  Drop-down list boxes
  Icons
  List boxes
  Option buttons
  Options
  Pop-up menus
  Text boxes

Bad	Good
Installation pop-up menu, 2	—
InstallShield icon, 1	—

Keyboard shortcuts

See also

For related guidelines, see also:

- *Procedures* on page 54

Read this

Keyboard shortcuts enable experienced (or "power") users to execute program commands from the keyboard rather than with the mouse. Typing in commands from the keyboard is, for many users, quicker and easier than manipulating a mouse. For this reason, keyboard shortcuts are a special feature of many programs.

Do this

When indexing keyboard shortcuts, follow these guidelines:

- **Index by name and type**
  Index shortcuts by name and type, where type is "keyboard shortcuts."

Bad	Good

Ctrl-Alt-Del, 24
Ctrl-P, 27
	Ctrl-Alt-Del, 24
	Ctrl-P, 27
	keyboard shortcuts
	   Ctrl-Alt-Del, 24
	   Ctrl-P, 27

- **Index related procedures**

  Index procedures associated with keyboard shortcuts by verb and noun.

Bad	Good
creating keyboard shortcuts, 10–11	creating keyboard shortcuts, 10–11 keyboard shortcuts, creating, 10–11

Master indexing

See also

For related guidelines, see also:

- *Nesting* on page 46
- *Page ranges* on page 49

Read this

Master indexes are what result when you combine two or more indexes. Normally you create a master index only for books that are sold together as a reference set. These reference set indexes are extremely valuable to users who are looking for an answer to a specific question but don't know which book in the reference set contains the answer.

Standardizing individual indexes

Although master indexes are a bit trickier to develop than normal book indexes, they are not as complicated as you might think. Essentially, what you need to do is set guidelines for developing the individual book indexes. The better your guidelines, the more seamlessly the individual indexes fit together in the end. When the individual indexes are complete, you combine them and make any necessary adjustments.

Preparing individual indexes

Before compiling your master index, you need to modify individual indexes in the following ways:

- **Abbreviate book titles in page numbers**

  In a master index, page numbers alone are meaningless. Page numbers need to be associated with book titles. The easiest way to associate page numbers with book titles is to include abbreviated book titles in the page numbers themselves.

- **Spell out abbreviations in page footers**

  Page numbers with abbreviated book titles are cryptic at best unless you explicitly spell out the abbreviations for users. The easiest way to spell out the abbreviations is to include a key in the footer of every page of the master index.

- **Add primary product names**

  If you know that your individual indexes focus on distinct products, you will need to add primary product names to those individual indexes after you compile them into a master index. If you do not do so, master index users will have no way of knowing which product is associated with which entry.

Editing the master index

After you compile individual indexes into a master index, you will need to edit the master index, renesting entries to eliminate redundancies and inconsistencies. The more consistently the individual indexes are nested, the less you will have to renest entries in the master index.

Do this

When creating a master index, follow these guidelines, in order:

1 **Spell out abbreviations in page footers**

  To make clear what the abbreviated book titles in page numbers mean, include a key in the footer of each page of the master index that explicitly spells out each book title.

  Page footer

LAG	=	*AcmePro for Linux Administrator's Guide*
LCR	=	*AcmePro for Linux Command Reference*
LUG	=	*AcmePro for Linux User's Guide*
UAG	=	*AcmePro for UNIX Administrator's Guide*
UCR	=	*AcmePro for UNIX Command Reference*
UUG	=	*AcmePro for UNIX User's Guide*

2 **Abbreviate book titles in page numbers**

Before compiling individual indexes into a master index, include abbreviated book titles in the page numbers of individual indexes.

Book titles	Page numbers
AcmePro for Linux Administrator's Guide	installing from network, LAG-3 uninstalling from network, LAG-24 to LAG-26
AcmePro for Linux Command Reference	CREATE command, LCR-25 DELETE command, LCR-67
AcmePro for Linux User's Guide	creating user profile, LUG-6 modifying preferences, LUG-14 to LUG-27
AcmePro for UNIX Administrator's Guide	installing from network, UAG-4 uninstalling from network, UAG-27 to UAG-28
AcmePro for UNIX Command Reference	CREATE command, UCR-26 DELETE command, UCR-69
AcmePro for UNIX User's Guide	creating user profile, UUG-6 modifying preferences, UUG-14 to UUG-27

3   **Add primary product names**
    After you compile individual indexes into the master index, add
    primary product names to the master index.

Before	After

installing from network, LAG-3
uninstalling from network, LAG-24
   to LAG-26
	installing from network
	   Linux, LAG-3
	uninstalling from network
	   Linux, LAG-24 to LAG-26

installing from network, UAG-4
uninstalling from network, UAG-27
   to UAG-28
	installing from network
	   UNIX, UAG-4
	uninstalling from network
	   UNIX, UAG-27 to UAG-28

CREATE command, LCR-25
DELETE command, LCR-67
	CREATE command
	   Linux, LCR-25
	DELETE command
	   Linux, LCR-67

CREATE command, UCR-26
DELETE command, UCR-69
	CREATE command
	   UNIX, UCR-26
	DELETE command
	   UNIX, UCR-69

creating user profile, LUG-6
modifying preferences,
   LUG-14 to LUG-27
	creating user profile
	   Linux, LUG-6
	modifying preferences
	   Linux, LUG-14 to LUG-27

creating user profile, UUG-6
modifying preferences,
   UUG-14 to UUG-27
	creating user profile
	   UNIX, UUG-6
	modifying preferences
	   UNIX, UUG-14 to UUG-27

4 **Renest entries**

After you add primary product names to the master index, renest entries to eliminate redundancies and inconsistencies.

Before	After

installing from network
    Linux, LAG-3
installing from network
    UNIX, UAG-4
uninstalling from network
    Linux, AG-24 to AG-26
uninstalling from network
    UNIX, UAG-27 to UAG-28
	installing from network
	    Linux, LAG-3
	    UNIX, UAG-4
	uninstalling from network
	    Linux, AG-24 to AG-26
	    UNIX, UAG-27 to UAG-28

CREATE command
    Linux, LCR-25
CREATE command
    UNIX, UCR-26
DELETE command
    Linux, LCR-67
DELETE command
    UNIX, UCR-69
	CREATE command
	    Linux, LCR-25
	    UNIX, UCR-26
	DELETE command
	    Linux, LCR-67
	    UNIX, UCR-69

creating user profile
    Linux, LUG-6
creating user profile
    UNIX, UUG-7
modifying preferences
    Linux, LUG-14 to LUG-27
modifying preferences
    UNIX, UUG-15 to LUG-28
	creating user profile
	    Linux, LUG-6
	    UNIX, UUG-7
	modifying preferences
	    Linux, LUG-14 to LUG-27
	    UNIX, UUG-15 to UUG-28

Nesting

See also

For related guidelines, see also:

- *Master indexing* on page 41
- *Page ranges* on page 49
- *Procedures* on page 54
- *"See" references* on page 63
- *Sorting* on page 72
- *Topics* on page 82

Read this

Nesting is organizing index entries into hierarchies with two and sometimes even three levels.

Example:

```
creating
    files, 2
    folders, 3
```

In this two-level example, "files" and "folders" are nested under "creating." The meaning is clear. "Creating files" is on page 2. "Creating folders" is on page 3.

Nested entries visually signal the relative importance of index entries and subentries. By sending these visual (pre-verbal) signals to users, nested entries enable users to scan an index and pinpoint the particular information they want quickly and easily.

The virtues of intelligent nesting cannot be overemphasized. If speed is the measure of an index, nesting is the accelerator. The better you nest your index, the quicker users will find what they need. Remember, users don't read indexes. They use them as a last resort in moments of desperation, anger, and despair. Your *only* goal as an indexer is to get users the information they need instantly.

Do this

When creating any index entry, follow these guidelines:

- **Nest redundant entries**

  If two or more entries begin with the same first word or phrase, nest the entries under that common word or phrase.

Bad	Good

creating files, 2
creating folders, 3
	creating
	    files, 2
	    folders, 3
	files, creating, 2
	folders, creating, 3

scanning drives, Macintosh, 5
scanning drives, Windows, 7
scanning files, Macintosh, 6
scanning files, Windows, 8
	drives, scanning
	    Macintosh, 5
	    Windows, 7
	files, scanning
	    Macintosh, 6
	    Windows, 8
	Macintosh, scanning
	    drives, 5
	    files, 6
	scanning
	    drives
	        Macintosh, 5
	        Windows, 7
	    files
	        Macintosh, 6
	        Windows, 8
	Windows, scanning
	    drives, 7
	    files, 8

- **Avoid single subentries**

  Never nest a single entry under a word or phrase.

Bad	Good
system messages, 137	system messages, 137

- **Do not nest proper nouns**

  Do not break apart proper nouns (for example, product names).

Bad	Good

Netscape
      Messenger, 25
      Navigator, 36
	Netscape Messenger, 25
	Netscape Navigator, 36

Page ranges

See also

For related guidelines, see also:

- *Master indexing* on page 41
- *Nesting* on page 46

Read this

Page ranges are the first and last pages of topics. If a single topic extends over more than one page, you should indicate on which page the topic begins and on which page the topic ends (for example, "hard drive, 11–15"). By including a page range, you give users a valuable clue about the length (and hopefully depth) of the topic.

Nesting subtopics

If a topic extends over more than a few pages, nest subtopics under the main topic. In this instance, do *not* indicate a page range for the main topic. The range is already visible to users in the subtopics. Also, by not indicating a page range for the main topic, you enable yourself to later add subtopics not included under the main topic you originally indexed.

WARNING Never list more than one page range for a particular topic. More than one page range simply tells users—and anybody else who is awake—that you, the indexer, were too lazy to hunt down and index subtopics. In effect, you are publicly forcing users to do your job. This is not a good career move.

Types of topics

Distinguish between extended and adjacent topics:

- **Extended topics (page range)**
  If a topic extends over more than one page, use a page range for the index entry (for example, "printing, 2–3").

- **Adjacent topics (separate page numbers)**
  If two separate discussions of the same topic happen to appear on two adjacent pages, use consecutive (separate) page numbers for the index entry (for example, "printing, 2, 3").

TIP Consecutive (separate) page numbers can indicate that a topic is poorly organized in a book. In other words, the index has uncovered structural problems in the book. If you have any control over the structure of the book, fix the offending topic, then re-index the topic.

Punctuating page ranges

Punctuate page ranges as follows:

- **Hyphenated page numbers**
  Sometimes writers hyphenate page numbers to remind users which chapter they are reading, or to make it easier for the writing team to reorganize the book in a later release. If page numbers are hyphenated, use "to" to indicate page ranges (for example, "printing, 2–1 to 2–3"). Never use em dashes (—) to indicate page ranges (for example, "printing, 2–1 — 2–3") for hyphenated page numbers. Such a construction is visually confusing.

- **Unhyphenated page numbers**
  If page numbers are not hyphenated, use en dashes (–) to indicate page ranges (for example, "printing, 2–3").

Do this

When indexing topics that extend over more than one page, follow these guidelines:

- **Use page ranges for multiple-page topics**
  If a topic extends over more than one page, use a page range for the index entry.

Bad	Good
creating user profiles, 15 creating user profiles, 16	creating user profiles, 15–16

- **Do not include page ranges for separate discussions**

  If two separate discussions of the same topic happen to appear on two adjacent pages, nest the discussions as subentries of the topic.

Bad	Good

printing, 22, 23	printing
	    books, 23
	    documents, 22

- **Use "to" with hyphenated page numbers**

  If page numbers are hyphenated, use the word "to" instead of an em dash (—) to indicate page ranges.

Bad	Good

restoring hard disk,
    2–11 — 2–12
	restoring hard disk,
	    2–11 to 2–12

- **Use en dashes with unhyphenated page numbers**

  If page numbers are not hyphenated, use en dashes (–) to indicate page ranges.

Bad	Good

restoring hard disk,
    11 to 12
	restoring hard disk,
	    11–12

Prepositions

See also

For related guidelines, see also:

- *Articles* on page 24

Read this

Prepositions are words that link nouns or pronouns to other sentence elements by:

- **Direction**
  Examples: *across, into, to, toward*

- **Figurative location**
  Examples: *against, for, with*

- **Physical location**
  Examples: *at, beside, between, by, in, on, through, under*

- **Time**
  Examples: *after, before, during, since, until*

Although prepositions play an active and useful role in most sentences, they are almost always counterproductive in index entries. Within an index, prepositions take up valuable space but rarely add any tangible benefit to users. Indexes answer specific questions for users with immediate, pressing questions (for example, "How do I recover the file I just deleted?"). Save your verbal eloquence for fireside readers, not users in the thick of a firefight.

Do this

When creating any index entry, follow this guideline:

- **Avoid prepositions**
  Do not use prepositions unless it is absolutely necessary.

Bad	**Good**
Windows	Windows
compatibility with, 23	compatibility, 23
metering applications in, 24	metering applications, 24
removing viruses	removing viruses
from Macintosh applications, 89	Macintosh, 89
from Windows applications, 87	Windows, 87

Procedures

See also

For related guidelines, see also:

- *Nesting* on page 46
- *Topics* on page 82

Read this

Procedures are step-by-step instructions that answer a single question:

- How?

They tell users how to do something.

Procedures are distinguished from topics, which explain everything else:

- Who
- What
- When
- Where
- Why

In technical publications, any text that is not a topic is a procedure, and vice versa.

Evaluating documents

Before you begin indexing procedures, decide whether the documents you are indexing are well structured or poorly structured:

- **Well-structured documents**
  If the document you are indexing contains headings that clearly answer specific questions, your job is half-completed before you start indexing. You can quickly identify procedures and begin indexing immediately.

- **Poorly structured documents**

  If the document you are indexing contains headings that do not clearly answer specific questions, the chances are extremely high that the document is poorly organized and poorly written. In this case, the road ahead will be rough. Rather than indexing immediately, you must first read through the text line by line, separating procedural from non-procedural information as you go. Once you have identified the procedures, you can begin indexing.

Syntax for procedure entries

Index procedures by noun and verb, using gerunds (that is, verbs ending in "ing") instead of the active form of the verb.

Examples:

Section Headings	Index entries
Printing	printing, 39–45
To print a document	document, printing, 39 printing document, 39

Do this

When indexing procedures, follow these guidelines:

- **Cross-index procedures**

  Index procedures by noun and verb.

Bad	Good
document printing, 15	document, printing, 15 printing document, 15

- **Use gerunds**

  Use gerunds (that is, verbs ending in "ing") instead of the active form of the verb.

Bad	Good
print document, 15	document, printing, 15 printing document, 15

- **Nest gerunds and nouns**

  Nest gerunds under nouns, and vice versa.

Bad	Good
AIX, 25-26, 49-50	AIX managed nodes
agent software, 47-54	configuring, 25-26
configuring, 25-30	installing agent software, 49-50
installing, 49-54	agent software
Linux, 27-28, 51-52	description, 47-48
Solaris, 29-30, 53-54	installing on managed nodes
	AIX, 49-50
	Linux, 51-52
	Solaris, 53-54
	configuring managed nodes
	AIX, 25-26
	Linux, 27-28
	Solaris, 29-30
	installing agent software
	AIX, 49-50
	Linux, 51-52
	Solaris, 53-54
	Linux managed nodes
	configuring, 27-28
	installing agent software, 51-52
	managed nodes. *See* AIX managed nodes; configuring managed nodes; installing agent software; Linux managed nodes; Solaris managed nodes
	Solaris managed nodes
	configuring, 29-30
	installing agent software, 53-54

- **Avoid "using"**

  Never use the word "using" when indexing user manuals. Everything in a user manual is used by the user.

Bad	Good
using printer, 32	printing, 32

Product names

See also

For related guidelines, see also:

- *Abbreviations* on page 20
- *"See" references* on page 63

Read this

Product names are trademarked names for products. For legal reasons, these names should always be indexed exactly as they have been trademarked.

Index product names according to how they are used in the book you are indexing:

- **Primary product**
  The primary product is the subject of the book you are indexing. For example, Adobe FrameMaker would be the primary product in a book entitled *Using Adobe FrameMaker*. Do not include the primary product name in your index. The name is implied in the title of the book itself.

- **Secondary products**
  Secondary products are products made by the manufacturer of the primary product. For example, Adobe Photoshop could be a secondary product in a book entitled *Using Adobe FrameMaker*. Include secondary product names in your index.

- **Third-party products**
  Third-party products are products made by manufacturers other than the manufacturer of the primary product. For example, Xerox DocuTech could be a third-party product in a book entitled *Using Adobe FrameMaker*. Include third-party product names in your index.

Do this

When indexing product names, follow these guidelines:

- **Use trademarked names**
  Index product names exactly as they are trademarked.

Bad	Good
Explorer, 2	Microsoft Internet Explorer, 2
WinWord, 3	Microsoft Word, 3

- **Index meaningful discussions**
  Do not index every mention of a product name. Index a product name as such only if the product is discussed at length. If you find two or more distinct discussions of the product, nest the discussions as subentries.

Bad	Good
Xerox DocuTech, 2, 5, 17, 43, 81	Xerox DocuTech, 17

HP LaserJet 5000, 4, 25,	HP LaserJet 5000
	    configuring, 25
	    installing, 4

- **Ignore primary products**
  Do not include the primary product name in your index. This name is the subject of the book itself and is implied throughout the index.

Bad	Good

Adobe FrameMaker
  generating
    book, 25
    cross-reference, 6
    index, 38
    PDF, 45
    table of contents, 29
	generating
	  book, 25
	  cross-reference, 6
	  index, 38
	  PDF, 45
	  table of contents, 29

- **Index secondary and third-party products**
  Include company names in secondary and third-party product names, even if the company name is not part of the trademarked name.

Bad	Good
LaserJet 5000, 25	HP LaserJet 5000, 25 LaserJet 5000. *See* HP LaserJet 5000
DocuTech, 17	DocuTech. *See* Xerox DocuTech Xerox DocuTech, 17

- **Cross-reference trademarked names**
  Include "see" references that point from commonly used but non-trademarked product names to trademarked product names.

Bad	Good
Adobe Acrobat Reader, 1	Acrobat Reader. *See* Adobe Acrobat Reader Adobe Acrobat Reader, 1 Reader. *See* Adobe Acrobat Reader
Microsoft Word, 2	Microsoft Word, 2 WinWord. *See* Microsoft Word

Scheduling

See also

For related guidelines, see also:

▪   *Tools* on page 80

Read this

When you index a document can affect the usability of your indexes and documents. Each phase of the documentation process offers distinct advantages and disadvantages to indexers.

TIP   The process of indexing can help you diagnose and correct structural problems in documentation. For this reason, professional editors routinely use the process of indexing to diagnose and correct structural problems within product documentation. For details, see *Procedures* on page 54 and *Topics* on page 82.

Why concurrent phases fail

In concurrent documentation phases, you index documents *during* the writing or editing phase of the documentation cycle:

▪   **During the writing phase**

   If you index documents as you write them, you can use the index to diagnose and to correct structural problems within the documents as they occur.

▪   **During the editing phase**

   If you index documents as you edit them, you can use the index to diagnose and correct structural problems within the documents before the documents go into final production.

All that glitters is not gold. Although they seem to provide you with a golden opportunity to use the indexing process to proactively edit your documents, concurrent documentation phases are actually a formula for failure.

Concurrent documentation phases do not work because they violate three cardinal rules of information development:

- Do one thing at a time.

- Let people do what they do best.

- Survive the project.

Concurrent phases force you to do too many things at once, prevent teamwork, and burn out indexers. Concurrent documentation phases will drive you crazy.

Why sequential phases succeed

In sequential documentation phases, you index documents *after* the writing or editing phase of the documentation cycle:

- **Between the writing and editing phases**

  If you index documents after you write them but before you edit them, you can focus on the usability of the indexes themselves. At your leisure, you can note structural problems within the documents to be addressed in the editing phase. This approach is ideal for documents that are published only once.

- **After the editing phase**

  If you index documents after you edit them, you can focus on the usability of the indexes themselves. At your leisure, you can note structural problems within the documents to be addressed in the next release of the documents. This approach is ideal for documents that are revised and published frequently.

Sequential documentation phases succeed because they:

- Enable you to focus on simple tasks.

- Encourage teamwork.

- Safeguard your sanity.

Setting indexing metrics

A central factor in calculating when to index a document is how long it takes to index the document.

When calculating how long it takes to index a document, base your estimates on the following:

- **Realistic experience**

  Always base indexing metrics on in-house experience, not numbers you read in a book. If you don't have any indexing experience, say so up-front, make a thumbnail guess, and upgrade your "guesstimate" to a realistic estimate once a clear pattern emerges. As with any estimate, add extra time for unexpected problems.

- **Page counts**

  Always base indexing metrics on the size of the document you are indexing, not on the size of the index itself. Remember that a well-structured (that is, well-nested) index takes up less space than a poorly structured index.

Do this

When deciding at which point in the documentation cycle to index your documents, follow these guidelines:

- **If your documents are published only once**

  If you have only one indexing cycle, make the most of it by indexing your documents before you edit them. This approach will enable you to diagnose and correct problems with the documents before they go into final production.

- **If your documents are revised frequently**

  If you have many indexing cycles, you can afford to index documents after you edit them. This approach will enable you to clean up the structure of your documents before you begin indexing, thereby simplifying the indexing process.

- **Base your schedule on reality**

  Base your indexing schedule on your worst in-house experience. Base estimates on the size of the document to be indexed, not on the index itself.

"See" references

See also

For related guidelines, see also:

- *Abbreviations* on page 20
- *Acronyms* on page 23
- *Capitalization* on page 28
- *Interface components* on page 36
- *Nesting* on page 46
- *Product names* on page 57
- *"See also" references* on page 67
- *System messages* on page 77
- *Topics* on page 82

Read this

A "see" reference is a cross-reference from an alias index entry to a preferred index entry.

Examples:

Alias entry	Preferred entry
small computer systems interface. *See* SCSI	SCSI, 25
dialog boxes. *See entries of specific dialog boxes*	Distribution Time dialog box, 2 Lock Package dialog box, 5

"See" references are extremely useful for users and indexers alike. They enable users to find items quickly, using whatever words come to mind during their search, however haphazard or hairbrained. At the same time, "see" references enable you, the indexer, to index items only once, thereby avoiding redundant work. "See" references enable you to structure the index logically, and they enable users to search the index illogically, yet to find exactly what they want. You can create an unlimited number of "see" references.

There are times when you may want to "double-post" the same item under multiple index entries with identical page numbers (for example, by listing procedures under nouns and gerunds, as this guide recommends). Some indexers actually advocate such redundancy for all entries because it gives users exactly what they want wherever they are in the index. Although there is some validity to this argument, multiple entries for the same item will increase the size of your index exponentially. In contrast, "see" references help streamline your index, making it easier for users to locate unique information, and making it easier for you to update the index when your document is updated.

TIP To create effective "see" references, turn your back on your company culture and go native. Walk in the shoes of your users. Look at the world from their perspective. Speak their language. Localize your index to the idiosyncrasies of your users. Now is the time to get creative. Create as many "see" references as you can possibly think of.

Do this

When developing "see" references, follow these guidelines:

- **Point to preferred items**
  Use "see" references to point from alias entries to preferred entries.

Bad	Good
exit, 5	exit, 5
quit, 5	quit. *See* exit

- **Point to abbreviations and acronyms**
  Use "see" references to point from complete words and phrases to their abbreviations or acronyms.

Bad	Good
small computer systems interface (SCSI), 13	SCSI, 13 small computer systems interface. *See* SCSI

- **Point to primary entries**
  Attach "see" references only to primary entries that have no page numbers or secondary entries.

Bad	Good

deleting
   files, 4
   folders, 6
erasing. *See* deleting
   files, 4
   folders, 6
	deleting
	   files, 4
	   folders, 6
	erasing. *See* deleting

- **Point to interface components**
  Use "see" references to point to specific interface components. For details, see *Interface components* on page 36.

Bad	Good

dialog boxes
   Distribution Time
      dialog box, 2
   Lock Package dialog
      box, 4
Distribution Time dialog
   box, 2
Lock Package dialog
   box, 4
	dialog boxes. *See entries of specific dialog boxes*
	Distribution Time dialog
	   box, 2
	Lock Package dialog
	   box, 4

- **Punctuate with periods**

  Use period (.) delimiters to separate "see" references from main entries.

Bad	Good
exit, 5	exit, 5
quit, *See* exit	quit. *See* exit

"See also" references

See also

For related guidelines, see also:

- *Cross-references* on page 31
- *"See" references* on page 63

Read this

A "see also" reference is a cross-reference from one index entry to a *related* index entry.

Examples:

Entry 1	Entry 2

installing
    See also de-installing
    DEC Alpha NT, 13-14
    Linux, 15-16
	de-installing
	    See also installing
	    DEC Alpha NT, 17-18
	    Linux, 19-20

hyperlinks, 55
    See also online help
	online help, 4
	    See also hyperlinks

"See also" references are distinguished by two qualities:

- **Independence**
  Each entry is independent of the other. Each would be valid if the other did not exist.

- **Mutual relevance**
  Each entry is somehow related to the other. It is highly probable that users who see one entry would like to see the other entry also. "See also" references always appear in reciprocal pairs.

> **TIP** To create effective "see also" references, put yourself in the shoes of users who are trying to learn more about the product being documented. For example, users who are interested in installing the product may well be interested in de-installing it. Do them a favor and make the connection explicit.

Do this

When developing "see also" references, follow these guidelines:

- **Point to related entries**

  Use "see also" references to point from index entries to related index entries.

Bad	Good
configuring managed nodes	configuring managed nodes
AIX, 25-26	*See also* installing agent
Linux, 27-28	software
Solaris, 29-30	AIX, 25-26
installing agent software	Linux, 27-28
AIX, 49-50	Solaris, 29-30
Linux, 51-52	installing agent software
Solaris, 53-54	*See also* configuring managed
	nodes
	AIX, 49-50
	Linux, 51-52
	Solaris, 53-54

- **Point to primary entries**

  Attach "see also" references to primary (first-level) entries only.

Bad	Good

configuring managed nodes
    AIX, 25-26. *See also* installing
        AIX
    Linux, 27-28. *See also* installing
        Linux
    Solaris, 29-30. *See also*
        installing Solaris
installing agent software
    See also configuring AIX, Linux,
        and Solaris
    AIX, 49-50
    Linux, 51-52
    Solaris, 53-54
	configuring managed nodes
	    See also installing agent
	        software
	    AIX, 25-26
	    Linux, 27-28
	    Solaris, 29-30
	installing agent software
	    See also configuring managed
	        nodes
	    AIX, 49-50
	    Linux, 51-52
	    Solaris, 53-54

- **Point in both directions**

  Use "see also" references in reciprocal pairs.

Bad	Good

installing
    AIX, 25-26
    Linux, 27-28
    Solaris, 29-30
requirements
    See also installing
    AIX, 19-20
    Linux, 21-22
    Solaris, 23-24
	installing
	    See also requirements
	    AIX, 25-26
	    Linux, 27-28
	    Solaris, 29-30
	requirements
	    See also installing
	    AIX, 19-20
	    Linux, 21-22
	    Solaris, 23-24

- **Separate with semicolons**

  Use semicolons (;) to separate "see also" references to multiple items.

Bad	Good

hyperlinks, 55
    See also online documentation
        and online help
online documentation, 4
    See also hyperlinks and online
        help
online help, 22
    See also hyperlinks and online
        documentation
	hyperlinks, 55
	    See also online
	        documentation;
	        online help
	online documentation, 4
	    See also hyperlinks; online help
	online help, 22
	    See also hyperlinks; online
	        documentation

- **Sort alphanumerically**

  List multiple "see also" references alphanumerically (that is, alphabetically and numerically).

Bad	Good

hyperlinks, 55
    See also online help;
        online
        documentation
online documentation, 4
    See also online help; hyperlinks
online help, 22
    See also online
        documentation;
        hyperlinks
	hyperlinks, 55
	    See also online
	        documentation;
	        online help
	online documentation, 4
	    See also hyperlinks; online help
	online help, 22
	    See also hyperlinks; online
	        documentation

- **Position at top**

  If the entry you are cross-referencing has subentries, list "see also" references as the first second-level entry.

Bad	Good
installing	installing
AIX, 25-26	*See also* requirements
Linux, 27-28	AIX, 25-26
See also requirements	Linux, 27-28
Solaris, 29-30	Solaris, 29-30
requirements	requirements
AIX, 19-20	*See also* installing
Linux, 21-22	AIX, 19-20
See also installing	Linux, 21-22
Solaris, 23-24	Solaris, 23-24

Sorting

See also

For related guidelines, see also:

- *Nesting* on page 46

Read this

Index entries are sorted alphabetically. That sounds simple enough. Entries beginning with "a" are listed under "A," entries beginning with "b" are listed under "B," and so on. However, when you start building an index, you will find that sorting is a bit more complicated than it first appears.

Sorting alphabetically

To begin with, there are three different ways to sort alphabetically:

- **By capitalization (machine)**

  Machines sort capital letters before lowercase letters. For example, they place "Apple" before "access" even though "c" comes before "p" in the alphabet. People, of course, do just the opposite, placing "access" before "Apple."

- **By letter (machine)**

  Machines sort words by letter, ignoring blank spaces between words. For example, they sort "Ice Cream" before "I scream" because they recognize only the letters "IceCream" and "Iscream."

- **By word (human)**

  People sort words by word, not letter. For example, you would sort "I scream" before "Ice Cream" because you recognize each whole word.

Sorting numbers

To complicate matters further, indexes may contain entries that begin with numbers.

There are two different ways to index numeric entries:

- **By digit (machine)**

  Machines sort numbers digitally. For example, they place "10" before "8" because the digit 1 comes before the digit 8.

- **By value (human)**

  People sort numbers numerically. For example, they place "8" before "10" because the number 8 is smaller than the number 10.

Sorting symbols

To complicate matters even further, indexes may contain entries that begin with symbols.

There are two different ways to sort symbolic entries:

- **By symbol (machine)**

  Machines sort entries beginning with symbols by symbol alone. For example, they place ".BAT files" after ". (period)" rather than after "backup."

- **By meaning (human)**

  People sort entries beginning with symbols by meaning. For example, they place ".BAT files" after "backup" rather than after ". (period)."

Do this

When sorting index entries, follow these guidelines:

- **Sort words for humans**

  Sort word by word, not letter by letter.

Bad	Good
Acme Computing, 37 aborting print jobs, 34	aborting print jobs, 34 Acme Computing, 37
printer fonts, 22 print version, 57	print version, 57 printer fonts, 22

- **Sort numbers for humans**

  Sort numbers by value, not digit.

Bad	Good
10-key calculator, 5 8 mm tape drives, 26	8 mm tape drives, 26 10-key calculator, 5

- **Sort symbols for humans**

  Sort symbols by meaning, not symbol.

Bad	Good
Symbols : (colon), 7 .BAT files, 8	**Symbols** : (colon), 7 **B** .BAT files, 8

- **Separate symbols, numbers, and words**
  Index symbolic, numeric, and alphabetic entries under separate headings.

Bad	Good
: (colon), 7	**Symbols**
.BAT files, 8	: (colon), 7
10-key calculator, 5	
8 mm tape drives, 26	**Numbers**
Acme Computing, 37	8 mm tape drives, 26
aborting print jobs, 34	10-key calculator, 5
	A
	aborting print jobs, 34
	Acme Computing, 37
	B
	.BAT files, 8

- **Index symbols with spelled-out forms in parentheses**
  Index symbols such as colons (:) and semicolons (;) by symbol, with their spelled-out forms in parentheses.

Bad	Good
B	**Symbols**
colon (:), 7	; (semicolon), 8
semicolon (;), 8	: (colon), 7

- **Index dot suffixes as alphabetic entries**

  Index names or suffixes beginning with a dot (.) as alphabetic rather than symbolic entries.

Bad	Good
Symbols .BAT files, 8 .DOC files, 9	**B** .BAT files, 8 **D** .DOC files, 9

System messages

See also

For related guidelines, see also:

- *Interface components* on page 36
- *"See" references* on page 63

Read this

System messages are (or should be) plain-language words, phrases, and sentences delivered by a computer program to users in particular situations. These messages are an invaluable part of the conversation between user and program. Unfortunately, they are also the most poorly designed components of most computer programs.

Types of system messages

There are three types of system messages:

- **Error messages (most common)**
  Error messages appear when something goes wrong. These negative messages are very common and often cryptic (for example, "404"). Many are built into tools programmers use to develop programs. These messages should never be seen by users.

- **Warnings (less common)**
  Warnings appear when something could go wrong. These proactive messages are found in user-friendly program interfaces and are usually well written (for example, "Are you sure you want to delete the file?"). They are, of course, much less common than error messages.

- **Success messages (rare)**
  Success messages appear when users perform an action successfully (for example, "Your file was uploaded successfully!"). These congratulatory messages are found only in program interfaces designed by people who take the user experience of the program seriously. They are extremely rare.

Quality of system messages

The quality of system messages varies widely. It ranges from inhuman ("404") to truly useful ("Are you sure you want to delete the file?") to overly friendly ("Have a nice day!"). Nevertheless, all system messages have one thing in common: they are either too cryptic or too long-winded to be indexed individually, word for word. Therefore, you should index them under "system messages" and cross-reference them by type (for example, "error messages. *See* system messages").

Do this

When indexing system messages, follow these guidelines:

- **Index system messages as a group**

   Do not index system messages individually. Index all system messages under "system messages." Include subentries where appropriate.

Bad	Good
404, 37	404. *See* system messages
file not found, 37	error messages. *See* system messages
folders have disappeared, 86	file messages. *See* system messages
	folder messages. *See* system messages
	messages. *See* system messages
	system messages
	404, 37
	files, 37
	folders, 86

- **Cross-reference system message entries**

  Add "see" references for other words that might be used to describe system messages.

Bad	Good

system messages
    files, 37
    folders, 86
	error messages. *See* system messages
	messages. *See* system messages
	system messages
	    files, 37
	    folders, 86

- **Index all messages as system messages**

  If there is more than one type of system message, index the types under "system messages."

Bad	Good

Are you sure you want to delete the file?, 12
file not found, 37
	error messages. *See* system messages
	messages. *See* system messages
	system messages
	    error messages, 37
	    warnings, 12
	warnings. *See* system messages

- **Ignore success messages**

  Do not index success messages. Success does not need to be explained, let alone indexed.

Bad	Good
File uploaded successfully!, 29	—

Tools

See also

For related guidelines, see also:

- *Scheduling* on page 60

Read this

Which indexing tools you use can affect both the usability and the reusability of your indexes. Each type of tool offers distinct advantages and disadvantages to indexers.

In very general terms, there are two types of indexing tools:

- **Internal tools**

  If you use an indexing tool that is integrated into a program (for example, Adobe FrameMaker), you can create index markers, or tags, that are embedded in the documents you are indexing. Internal tools are ideal for indexing documents that are revised frequently. For example, if you add a chapter at the beginning of a book, the page numbers for index entries in subsequent chapters are updated automatically the next time you generate the index. The disadvantage of these tools is that you normally need the exact version and platform of the software used to create the documents you are indexing.

- **External tools**

  If you use a stand-alone indexing tool (for example, CINDEX), you can create an index that is independent of the platform and program of the document you are indexing. External tools are ideal for indexing hardcopy documents or electronic documents that are not revised frequently. As a rule, these tools sort entries automatically and include special menus designed exclusively for indexing functions. The disadvantage of these tools is that you enter page numbers manually. If the pagination of the document changes, you must update these page numbers by hand.

Do this

When choosing an indexing tool, follow these guidelines:

- **Choose a flexible indexing method**

  Choosing an indexing tool is a classic chicken-and-egg problem. Before you can test a tool, you need to decide what you want it to do. That is, you need an indexing approach, or method. But before you can implement your chosen indexing method, you need a tool. The sequential indexing method presented in this guide is program- and platform-independent. Adapt this method to the particular needs of your organization, products, and documents. Then road-test your adapted indexing method using whatever indexing, desktop publishing, or word processing tools you already have in-house. Once you have tested and refined your indexing method, look for tools that help you automate your method. At that point, you can pick a tool that exactly meets your needs.

- **Choose a flexible indexing tool**

  Although there are excellent arguments for using currently available internal and external indexing tools, it is unwise to fixate on any one tool or technology. Information development and management technologies are evolving at breakneck pace. Yet, as the Internet so publicly and painfully demonstrates, machines alone cannot debabelize complex information structures. Machines parse data, not information. They have strong backs but weak minds. No tool can "solve" the problem of indexing. At best, tools help you automate a solution of your own making. Choose an indexing tool that maximizes your flexibility. Use the tool to automate the indexing method you have already tested, implemented, and refined in-house.

- **Choose a local solution**

  When evaluating indexing tools, avoid solutions to other people's problems. You have enough problems of your own. Collect trial versions of the most flexible tools on the market and test them mercilessly on the biggest, ugliest in-house documents you can find. Whichever tool survives your local gauntlet is the tool for you.

Topics

See also

For related guidelines, see also:

- *Nesting* on page 46
- *Procedures* on page 54
- *"See" references* on page 63

Read this

Topics are narrative text that answers one of the following questions:

- Who?
- What?
- When?
- Where?
- Why?

Topics are distinguished from procedures, which always answer a single question:

- How?

In technical publications, any text that is not a procedure is a topic, and vice versa.

Indexing well-structured topics

In the same way that well-structured user interfaces are easy to document, well-structured topics are easy to index. To find out if the topics you are indexing are well structured, look first at their headings.

Well-structured headings answer specific questions.

Examples:

Question	Heading
What is Norton AntiVirus?	About Norton AntiVirus
What does Norton AntiVirus do?	How Norton AntiVirus works
Why should I use Norton AntiVirus?	Why use Norton AntiVirus
When should I scan drives?	When to scan drives
What types of viruses exist?	Types of viruses
What do I do if AutoScan doesn't start?	If AutoScan doesn't start
How does printing work?	Printing
How do I print a document?	To print a document

Analyzing topics

Before you begin indexing topics, decide whether the document you are indexing is well structured or poorly structured:

- **Well-structured documents**

  If the document you are indexing contains headings that clearly answer specific questions, your job is half-completed before you start indexing. You can quickly identify topics and begin indexing immediately.

- **Poorly structured documents**

  If the document you are indexing contains headings that do not clearly answer specific questions, the document is probably poorly organized and poorly written. In this case, the road ahead will be rough. Rather than indexing immediately, you must first read through the text line by line, separating procedural from non-procedural information as you go. Once you have identified the (non-procedural) topics, you must ask yourself what question each piece of text answers. Then and only then can you begin indexing.

Syntax for topic entries

Indexing well-structured topics is straightforward:

- Begin entries with the subject of each question.
- End entries with a single word indicating the question answered (for example, who, what, when, or why).

For example, well-structured headings could appear in an index as follows:

Headings	Index entries

About Norton Utilities
How Norton Utilities works
Why use Norton Utilities
	Norton Utilities
	    about, 7
	    how it works, 9–10
	    why use, 8

When to scan drives	drives, when to scan, 13
Types of viruses	viruses, types, 27
If automatic protection fails	automatic protection failure, 89 failure, automatic protection, 89 protection failure, 89

Do this

When indexing topics, follow these guidelines:

- **Cross-index topics**
  Index topics by noun and noun modifier.

Bad	Good
unable to repair infected files, 25	files, irreparable, 25 infected files, irreparable, 25 irreparable files, 25

- **Begin entries with subjects**

  Begin entries with the subject of each question answered by the text.

Bad	Good
about command-line options, 32	command-line options, 32 options, command-line, 32

where to find the Emergency Boot
  Disk, 70
if the Emergency Boot Disk fails, 71
	Emergency Boot Disk
	  failure, 71
	  location, 70

- **End entries with predicates**

  End entries with a single word indicating the question answered (for example, who, what, when, or why).

Bad	Good

Norton AntiVirus, 7, 8, 9–10	Norton AntiVirus
	  about, 7
	  how it works, 9–10
	  why use, 8

drives, 13	drives, when to scan, 13
types of viruses, 27	viruses, types, 27

- **Create multiple entries**

  If an index entry contains more than one word that could be constructed as a noun, create new entries, or aliases, for each noun.

Bad	Good
command-line options, 32	command-line options, 32 options, command-line, 32
unable to repair infected files, 25	files, irreparable, 25 infected files, irreparable, 25 irreparable files, 25

- **Cross-reference topics**

  If an index entry contains subentries, use "see" references to point from the alias entries to the index entry that has subentries.

Bad	Good

privileges
    read, 40
    write, 41
	access rights. *See* privileges
	privileges
	    read, 40
	    write, 41
	read privileges. *See* privileges
	rights. *See* privileges
	write privileges. *See* privileges

INDEX

Symbols

– (en dash), 51
— (em dash), 51

A

abbreviating book titles
    description, 41
    guidelines, 43
abbreviations
    See also acronyms; product names
    cross-referencing, 65
    description, 20–21
    guidelines, 21–22
    imposing predictability, 21
    types, 20
acronyms
    See also abbreviations
    cross-referencing, 65
    description, 23
    guidelines, 23
active components
    description, 36
    guidelines, 37
adapting sequential method, 81
adjectives. *See* articles
advertising slogans, 25
alias entries
    See also preferred entries
    cross-referencing, 15
    description, 63
    guidelines, 64
alphabetical sorting. *See* sorting
analyzing topics. *See* topics

anticipating problems. *See* scheduling
appendices
    creating entries, 13
    description, 26
    guidelines, 27
articles
    See also prepositions
    avoiding, 25
    deadly to indexes, 24–25
    essential to sentences, 24
    removing, 14
automating sequential method. *See*
  sequential method
avoiding
    articles, 25
    company culture, 64
    downstyle capitalization, 28
    prepositions, 52–53
    redundancy
        abbreviations, 21
        master index, 45
        tasks, 15
    scheduling problems, 62
    single subentries, 48
    "using", 56

B

back matter, 26–27
book titles, 57
buttons
    description, 36
    guidelines, 37

C

calculating job metrics. *See* scheduling

capitalization

    See also capitalizing; sorting

    description, 28

    guidelines, 29–30

    visual signals, 28

capitalizing

    See also capitalization

    commands, 28

    filenames, 28

    interface components, 28, 29

    literal strings, 28

    menu items, 28

    program names, 28

catch-phrases, 25

chapters, indexing, 8

chicken-and-egg problem, 81

combination boxes

    description, 36

    guidelines, 38

commands

    See also menu commands; typed
       commands

    capitalizing, 28–29

    categorizing, 36–37

common abbreviations. *See* abbreviations

communication, visual, 28

company culture, avoiding, 64

comparisons

    active v. passive interface components,
       36

    adjacent v. extended topics, 49

    concurrent v. sequential indexing phases,
       60–61

    content v. format, 3

    definite v. indefinite articles, 24

    document size v. index size, 62

    em dash (—) v. "to" in page ranges, 50

    generic v. specific nouns, 24

    headings v. index entries, 55

    headings v. questions, 82

comparisons *(continued)*

    human v. machine

       indexing, 81

       sorting, 72–73

    information v. data, 81

    page ranges v. consecutive page
       numbers, 49

    procedures v. topics, 54

    "see" v. "see also" references, 31

    user search patterns v. index structures,
       63

    well-structured v. poorly structured
       documents, 83

complex tasks. *See* simplifying tasks

components. *See* product components

computer program. *See* product components;
   product names; products; tools

concurrent phases, 60–61

    See also sequential phases

content, creating, 3

continuing success. *See* success

copyright pages, 34–35

creating

    content, 3

    cross-references

       "see" references, 15

       "see also" references, 16

    entries

       back matter, 13

       chapters, 8

       front matter, 13

       procedures, 9

       product components, 12

       product names, 11

       topics, 10

    hierarchies, 46–48

    master index footers, 42

creativity. *See* "see" references

critical information. *See* reference information

cross-indexing

    See also cross-referencing

    procedures, 55

    topics, 84

cross-references
　　See also cross-referencing; "see"
　　　references; "see also" references
　　description, 31
　　guidelines, 32–33
cross-referencing
　　See also cross-indexing; cross-references;
　　　"see" references; "see also" references
　　abbreviations, 21–22
　　acronyms, 23
　　product names, 59
　　system messages, 78–79
　　topics, 86
customizing indexes, 64
customizing sequential method. *See*
　sequential method

D

definite articles. *See* articles
dialog boxes
　　description, 36
　　guidelines, 37
digit-by-digit. *See* sorting
disks
　　description, 36
　　guidelines, 37
document
　　conventions, 34–35
　　structure
　　　headings, 82–84
　　　procedures, 55
　　　topics, 83
documentation process
　　phases, 60–62
　　quality assurance, 4
dot (.) filename suffixes, 76
downstyle capitalization, 28
　　See also lowercase; uppercase
drop-down list boxes
　　description, 36
　　guidelines, 38

E

editing
　　individual indexes
　　　description, 3
　　　steps, 14
　　master indexes
　　　description, 42
　　　guidelines, 45
editing phase. *See* documentation process
editors, viii
　　See also indexers; users
em dash (—). *See* page numbers; page ranges
embedding index markers
　　revising, 80
　　testing, 17
emergency procedures
　　description, 26
　　guidelines, 27
en dash (–). *See* page numbers; page ranges
encouraging teamwork, 61
end users. *See* users
entries, meaningful, 58
entry levels
　　alias entries, 63–64
　　preferred entries, 63–64
　　primary entries
　　　"see" references, 65
　　　"see also" references, 69
　　related entries, 67–68
　　secondary entries
　　　"see" references, 65
　　　"see also" references, 71
error messages. *See* system messages
estimating times. *See* scheduling
evaluating
　　headings, 82–84
　　procedures, 54
　　topics, 83

F

families, product. *See* master indexes; reference sets
features, undocumented, 5
filenames
    capitalizing, 28–29
    description, 36
    guidelines, 37
footers, master index, 42
format, editing, 3
front matter
    description, 34
    guidelines, 35

G

generic components
    description, 36
    guidelines, 38
generic nouns, 24
gerunds
    See also verbs
    description, 55
    guidelines, 55
getting started, 7–17
glossaries
    description, 26
    guidelines, 27
"guesstimates", 62

H

headings, document, 82–84
help. *See* getting started
hierarchies
    description, 46
    guidelines, 47–48
    procedures, 9
    topics, 10
honest estimates, 62
hyphenated page numbers. *See* page numbers

I

icons
    description, 36
    guidelines, 38
implementing sequential method. *See* sequential method
increasing
    profits, 4–5
    speed, 46
indefinite articles. *See* articles
indexers
    See also editors; indexes; managers; users; writers
    novices, 2
    professionals, 2
indexes
    See also indexers
    creating, 8–13
    cross-referencing, 15–16
    customizing, 64
    editing, 14
    structuring logically, 63
    testing, 17
    usability, 4
information developers. *See* writers
initial capitalization
    description, 28
    guidelines, 29
interface components
    See also system messages
    capitalizing, 28, 29
    creating entries, 12
    cross-referencing, 65
    description, 36
    guidelines, 37–38
introductions
    creating entries, 13
    description, 34
    guidelines, 35

J

jobs, scheduling. *See* scheduling

K

keyboard shortcuts
    creating entries, 12
    description, 39
    guidelines, 39–40

L

left-brain test, 17
    See also right-brain test
letter-by-letter. *See* sorting
libraries. *See* master indexes; reference sets
linguistic shortcuts. *See* abbreviations;
  acronyms
list boxes
    description, 36
    guidelines, 38
literal strings, 28–29
localizing indexes. *See* customizing indexes
lowercase
    See also downstyle capitalization;
     uppercase
    description, 28
    guidelines, 29

M

managers, viii
    See also indexers; users
managing editors. *See* editors; managers
markers, index, 80
master indexes
    See also reference sets
    abbreviating book titles
        description, 41
        guidelines, 43
    adding product names
        description, 42
        guidelines, 44

master indexes *(continued)*
    creating footers, 42
    editing
        description, 42
        guidelines, 45
    renesting entries, 45
    standardizing, 41
meaning-by-meaning. *See* sorting
meaningful entries, 58
menu commands
    See also commands; menu items; typed
     commands
    description, 36
    guidelines, 37
menu items, 28
    See also menu commands; menus
menus
    See also menu items
    description, 36
    guidelines, 37
messages. *See* system messages
method, indexing. *See* sequential method
metrics, indexing, 61
mindset. *See* users
mouse. *See* keyboard shortcuts
multimedia, viii
multiple entries, 85

N

names. *See* product names
narrative text. *See* topics
nesting
    See also sorting
    gerunds, 56
    increasing speed, 46
    individual indexes
        description, 46
        guidelines, 47–48
        steps, 14
    master indexes, 45
    nouns, 56
    procedures, 56

nesting *(continued)*
    product names, 58
    proper nouns, 48
    single entries, 48
    subtopics, 49
    topics, 10
    verbs, 55
non-verbal communication. *See* visual
  communication
nouns
    generic, 24
    nesting
        description, 55
        guidelines, 56
    proper, 48
numbers, sorting. *See* sorting

O

online publications, viii
option buttons
    description, 36
    guidelines, 38
options
    description, 36
    guidelines, 38

P

page counts, 62
page numbers
    See also page ranges
    cross-references
        "see" references, 65
        "see also" references, 31
    hyphenated
        description, 50
        em dashes (—), 51
        guidelines, 51
        "to", 50
    master index
        abbreviated book titles, 43
        numbers not enough, 41

page numbers *(continued)*
    unhyphenated
        description, 50
        en dashes (–), 51
        guidelines, 51
page ranges
    See also page numbers
    description, 49–50
    em dashes (—), 51
    en dashes (–), 51
    guidelines, 50–51
    procedures, 9
    product names, 11
    punctuating, 50
    "to", 50
    topics, 10
painless indexing, 2
pairs, cross-reference. *See* "see also"
  references
parameters
    description, 36
    guidelines, 37
passive components
    description, 36
    guidelines, 37
period (.) delimiters. *See* "see" references
phases, documentation. *See* documentation
  process
platform-independence, 81
pop-up menus
    description, 36
    guidelines, 38
predicates, topic, 85
predictable abbreviations, 21
prefaces
    creating entries, 13
    description, 34
    guidelines, 35
preferred entries
    See also alias entries
    description, 63
    guidelines, 32

prepositions
See also articles
avoiding, 53
description, 52
guidelines, 53
removing, 14
pre-verbal communication. See visual
communication
primary entries
See also secondary entries
"see also" references, 69
"see" references, 65
primary product names. See product names
printed publications, viii
procedures
See also topics; troubleshooting
procedures
chapters, 8
creating entries, 9
cross-indexing, 55
description, 54–55
emergency, 26
gerunds, 55
guidelines, 55–56
nesting, 56
nouns, 56
subprocedures, 9
syntax, 55
troubleshooting, 26
verbs, 55
process, documentation. See documentation
process
product components
See also product names; products
capitalizing, 29
creating entries, 12
product families. See master indexes;
reference sets

product names
See also abbreviations; product
components; products
capitalizing, 29
master index
description, 42
guidelines, 44
nesting, 58
page ranges, 11
primary
creating entries, 11
description, 57
guidelines, 58
secondary
description, 57
guidelines, 59
third-party
description, 57
guidelines, 59
trademarked
cross-referencing, 59
description, 57
guidelines, 58
untrademarked, 59
products
See also product components; product
names
creating entries, 11
successful, 5
profits, increasing, 4–5
program-independence, 81
programs. See product components; product
names; products; tools
projects, surviving, 61
proper nouns. See nouns
publications managers. See managers

Q

quality assurance, 4

R

reality-based
    method, 3
    scheduling, 62
    tools, 81
reciprocal cross-references. *See* "see also"
  references
redundancy, avoiding
    abbreviations, 21
    master index, 45
    nesting, 47
    "using", 56
reference information
    back matter, 27
    front matter, 35
reference sets
    See also master indexes
    description, 41–42
    guidelines, 42–45
related entries
    description, 67
    guidelines, 68
requirements, system. *See* system
  requirements
revised documents, 61
right-brain test, 17
    See also left-brain test

S

sanity, preserving
    documentation phases, 61
    indexing steps, 3
scanning indexes, 25
scheduling
    description, 60–62
    guidelines, 62
searching indexes illogically. *See* users
secondary entries
    See also primary entries
    "see" references, 65
    "see also" references, 71
secondary product names. *See* product names

"see" references
    See also cross-references; cross-
      referencing; "see also" references
    abbreviations, 65
    acronyms, 65
    alias entries, 64
    creating, 15
    creativity, 64
    description, 63–64
    guidelines, 64–66
    interface components, 65
    page numbers, 31, 65
    period (.) delimiters, 66
    preferred entries, 64
    primary entries, 65
    product names, 59
    secondary entries, 65
    subentries, 32
    testing, 15
    topics, 86
    unlimited number, 64
    unreciprocal, 31
"see also" references
    See also cross-references; cross-
      referencing; "see" references
    creating, 16
    description, 67–68
    guidelines, 68–71
    independent entries, 67
    mutually relevent entries, 67
    page numbers, 31
    primary entries, 69
    reciprocal, 31
    related entries, 68
    secondary entries, 71
    semicolons (;), 70
    sorting, 70
    subentries, 71
    testing, 16
semicolons (;), 70
separating content and format. *See* sequential
  method

sequential method
    See also sequential phases; testing
    automating, 81
    customizing, 81
    description, 3
    implementing, 81
    success-based, vii
    testing, 81
sequential phases, 61
    See also concurrent phases; sequential
      method
shortcuts. *See* abbreviations; acronyms;
  keyboard shortcuts
signals, visual, 28
simplifying tasks
    sequential method, 3
    sequential phases, 61
single subentries. *See* subentries
size, index, 62
software. *See* product components; product
  names; products; tools
sorting
    See also capitalization; nesting
    alphabetically, 72
    articles, 25
    guidelines, 74–76
    human
        meaning-by-meaning, 73
        value-by-value, 73
        word-by-word, 72
    machine
        capitalization, 72
        digit-by-digit, 73
        letter-by-letter, 72
        symbol-by-symbol, 73
    numbers, 73
    "see also" references, 70
    symbols, 73
specific nouns, 24
speed, increasing, 46

standards
    corporate publications, viii
    master index, 41
starting index, 7–17
step-by-step method. *See* sequential method
strategy. *See* sequential method
strings, capitalizing, 28, 29
structuring
    cross-references, 63
    documentation, 4
    headings, 82–84
    indexes, 14
    procedures, 55
    topics, 83
students, technical communication, viii
subentries
    avoiding single, 48
    "see" references, 32
    "see also" references, 71
    system messages, 78
subjects, topic, 85
subprocedures, 9
subtopics, nesting, 49
success
    See also successful
    continuing, viii
    immediate, viii
success messages
    description, 77
    guidelines, 79
successful
    See also success
    documents, 4
    indexes, 4
    method, vii
    products, 5
    scheduling, 61
    tools, 81
surviving projects, 61
symbols, sorting. *See* sorting

syntax
    headings, 82
    procedures, 55
    topics, 84
system messages
    See also interface components
    back matter, 26–27
    categorizing, 79
    creating entries, 12
    cross-referencing, 78–79
    error messages, 77–78
    subentries, 78
    success messages, 79
    warnings, 77–78
system requirements
    description, 34
    guidelines, 35

T

tables of contents
    description, 34
    guidelines, 35
tactics. *See* sequential method
tags, index, 80
tasks, simplifying. *See* simplifying tasks
teachers, technical communication, viii
teamwork, encouraging, 61
technical editors. *See* editors
technical writers. *See* writers
testing
    See also sequential method
    cross-references
        "see" references, 15
        "see also" references, 16
    indexes, 17
    sequential method, 81
    tools, 81
text. *See* topics
text boxes
    description, 36
    guidelines, 38

third-party names. *See* product names
title pages
    description, 34
    guidelines, 35
titles, book, 57
"to". *See* page numbers; page ranges
tools
    description, 80
    flexibility, 81
    guidelines, 81
    local solutions, 81
    testing, 81
topics
    See also procedures
    adjacent, 50
    analyzing, 83
    back matter, 26
    chapters, 8
    cross-indexing, 84
    cross-referencing, 86
    description, 82–84
    extended, 49
    front matter, 35
    guidelines, 84–86
    headings, 82–84
    hierachies, 10
    indexing, 10
    nesting subtopics, 49
    page ranges, 10
    predicates, 85
    "see" references, 86
    subjects, 85
    syntax, 84
    well-structured, 82–84
trademarks
    abbreviations
        description, 20
        guidelines, 22
    capitalizing, 30
    product names
        cross-referencing, 59
        description, 57
        guidelines, 58

troubleshooting procedures
  See also procedures
  description, 26
  guidelines, 27
typed commands
  See also commands; menu commands
  description, 36
  guidelines, 37

U

uncommon abbreviations. *See* abbreviations
undocumented features, 5
unhyphenated page numbers. *See* page
  numbers
unreciprocal cross-references. *See* "see"
  references
untrademarked product names. *See* product
  names
uppercase
  See also downstyle capitalization;
    lowercase
  description, 28
  guidelines, 29
usability
  See also users
  documentation, 4
  improving, 61
  indexes, 4
  products, 5
users
  See also editors; indexers; managers;
    usability; writers
  as testers, 17
  attitude, 46
  idiosyncrasies, 64
  illogical searches, 63
"using" redundant, 56
utilities
  description, 36
  guidelines, 37

V

value-by-value. *See* sorting
verbs
  See also gerunds
  description, 55
  guidelines, 55
visual communication, 28

W

warnings. *See* system messages
windows
  description, 36
  guidelines, 37
word-by-word. *See* sorting
workload, reducing
  abbreviations, 21
  cross-references, 15
writers
  See also indexers; users
  as first users, 5
  as indexers, viii
writing phase. *See* documentation process